Atomic Physics for Kids

Basic & Advanced Concepts

Explained in Rhymes

Preetinder Rahil

The author may have checked for errors twice or even thrice, but doing your due diligence would still be considered wise.

Things that are big
are made of things that are small.
The fact you must
remember once and for all.

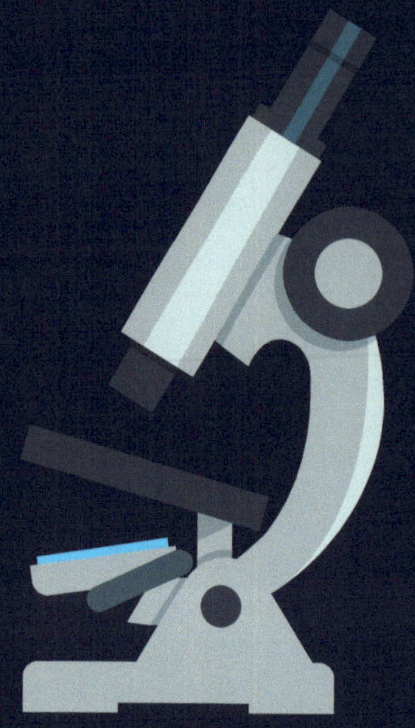

We are all made of atoms
that we can't see.
Looking under the electron microscope
is the key.

The centre is where the nucleus is found.
The electrons move around it,
round and round.

The nucleus is the atom's core.
Protons and neutrons are equal in number,
but in isotopes, they differ for sure.

Atomic number is labeled with the letter Z.
It's equal to the number of protons you see.

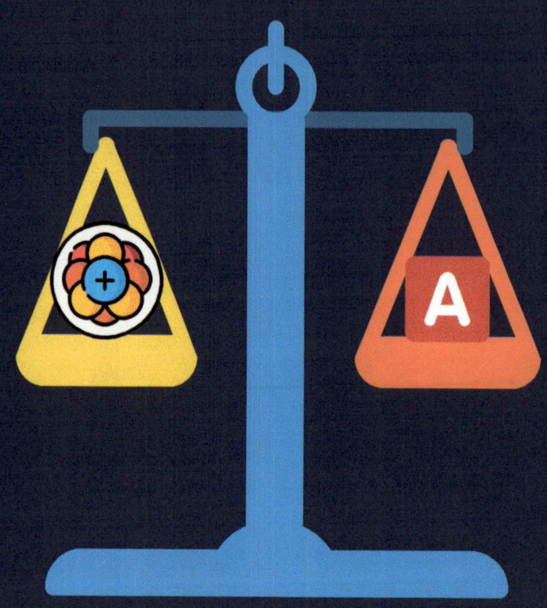

Atomic mass is labeled with the letter A.
Counting the number of protons and neutrons is the only way.

Protons and neutrons are made of heavy stuff.
Counting their masses to get the atomic mass
is more than enough.

When the number of neutrons in atoms differs,
but the number of protons stays the same.
We call them isotopes by name.

In the nucleus, the strong force is found.
It keeps protons and neutrons tightly bound.

If you really seek,
there's another nuclear force,
but it's quite weak.
An unstable nucleus radiates.
That's what the weak force mediates.

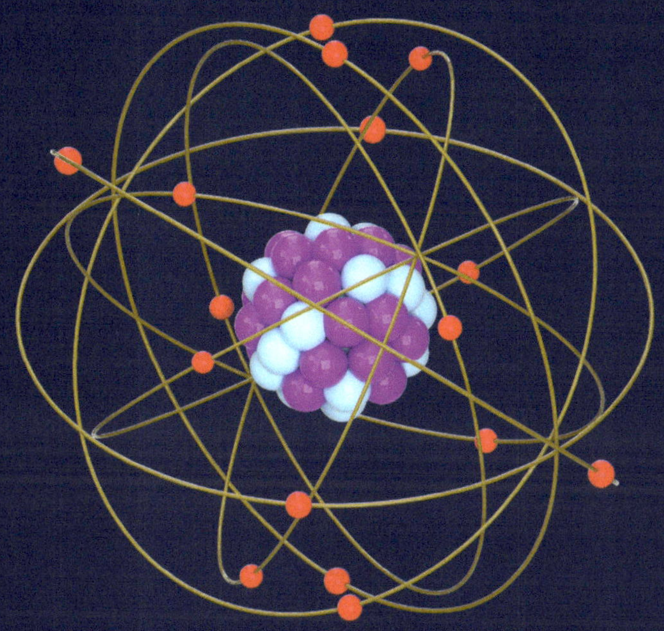

Protons and neutrons are made of quarks.
It's an experimental fact, not a guess.
I know more things mean more mess,
but from atomic physics, don't expect any less.

Quarks come in three colors.
Red, blue, and green.
They are not the real colors
which can be seen.
It's just a naming convention.
You know what I mean.

Quarks come in flavors too.
Up and down,
top and bottom,
charm and strange.
It's quite a range.

Proton has one down quark,
but its up quarks are two.
Thanks to the gluons,
they are held together like glue.

In the neutron town,
we have one up quark,
and two that are down.

Quarks are never found on their own.
In the theory, that's well known.
And in the experiments, that's been shown.

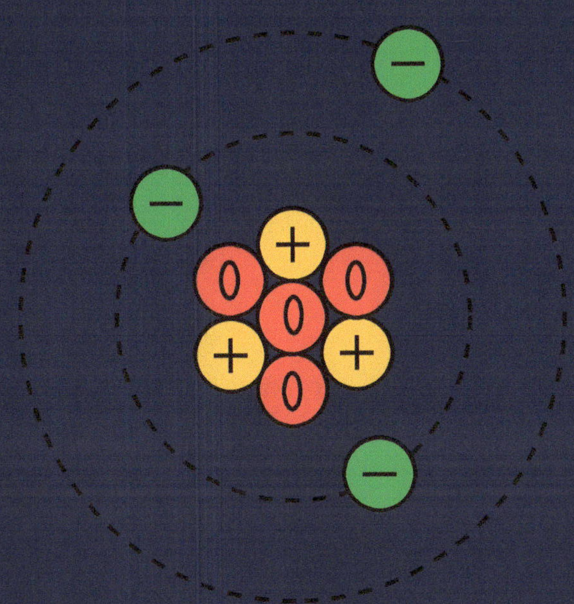

A proton has a charge of plus one.
An electron has a charge of minus one.
If you look for an electric charge
on a neutron, there's none.

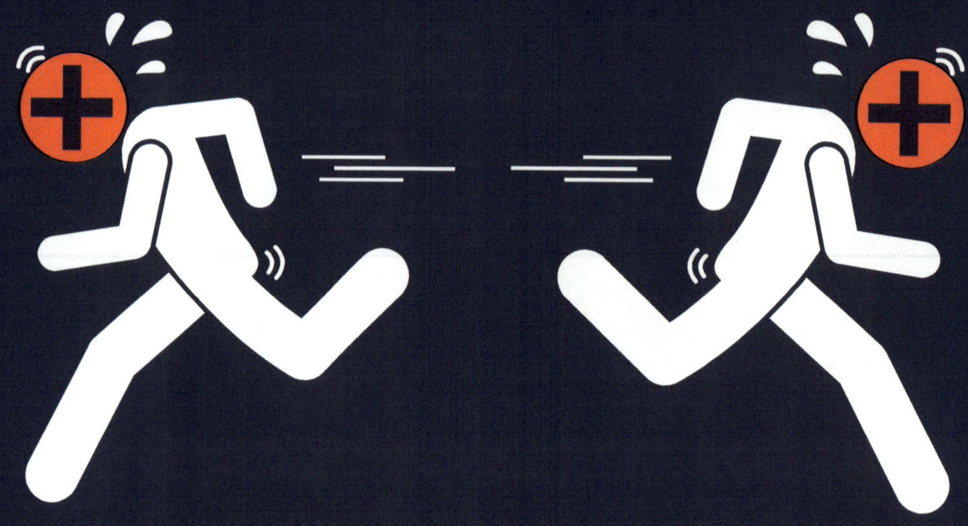

Like charges repel.
A force pushes them apart.
That's how you can tell.

Unlike charges attract.
A charged comb pulls on
bits of paper.
That's a fact.

Electrons revolve around.
Many orbitals are found.
The electron closest to the nucleus
is in a state called ground.

I must confess.

One thing I must address.

Things are not as simple as they sound.

Electrons are not like tiny balls that go around.

The old way of thinking we must shred.
Orbitals are probability waves that get spread.

What's the chance?
That's the probability at a glance.

If electrons don't go around,
where else can they be found?

You need a strong resolve.
As you have the formidable
Schrodinger equation to solve.

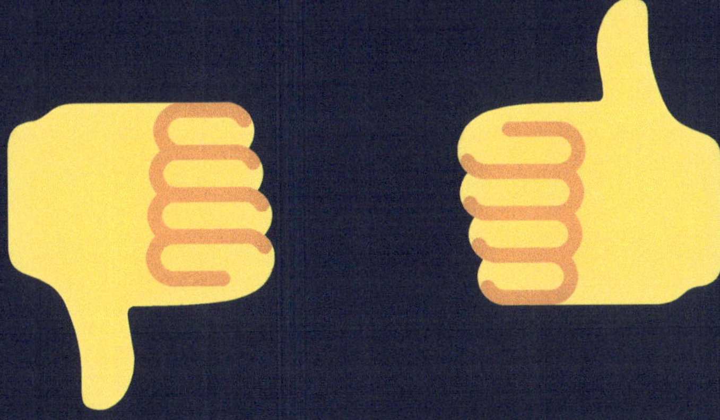

An electron has a spin.
It has only two values $+\frac{1}{2}$ and $-\frac{1}{2}$.
You get one or the other.
One you lose, the other you win.

You can't make the spin fast or slow.
Quantum spin is a fundamental property.
That much we know.

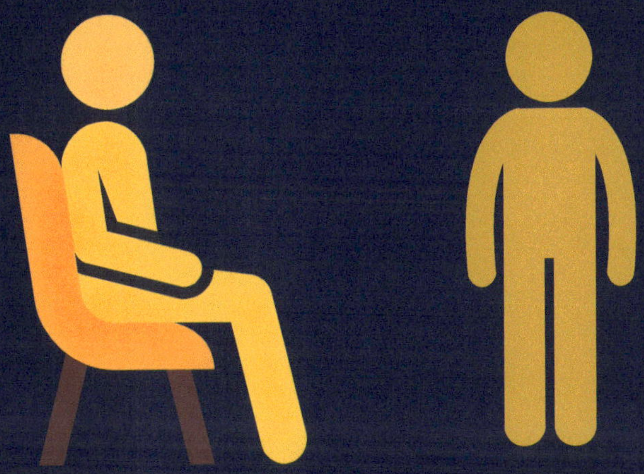

For electrons, the universe has chosen
a special fate.
They cannot occupy the same
quantum state.

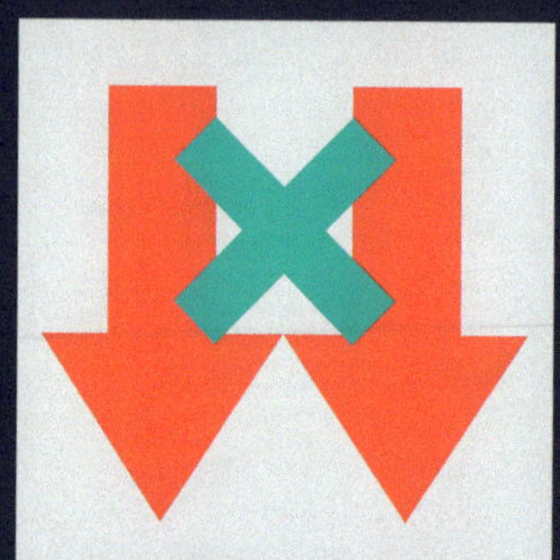

An orbital can have two electrons with opposite spins.
The fact that you must keep in mind.
That's how the Pauli exclusion principle is defined.

Are all atoms the same?

How can we tell them apart?

The answer lies in the periodic table chart.

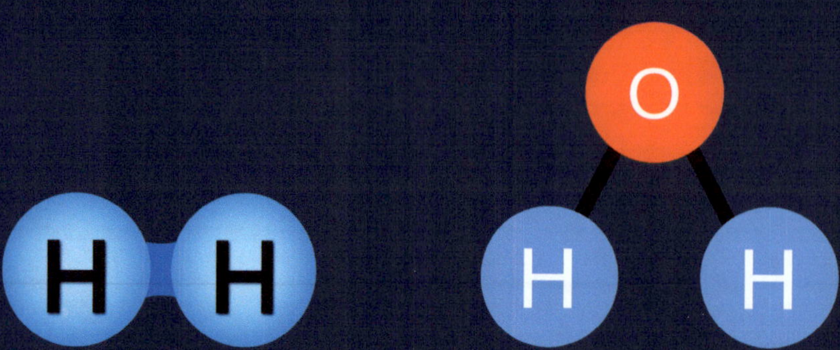

The periodic table is made of elements
which have atoms that are all the same.

Molecules too have atoms, but each one
maybe of a different name.

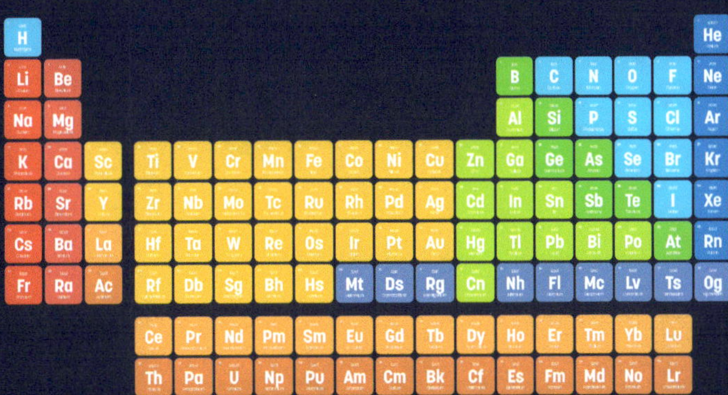

The periodic table is arranged in order
of atomic number and electrons
in the outermost shell.
This groups together the elements
as you can probably tell.

All fundamental forces and elementary
particles if we collect.
The standard model of particle physics
is what we get.

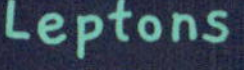

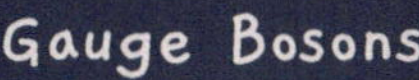

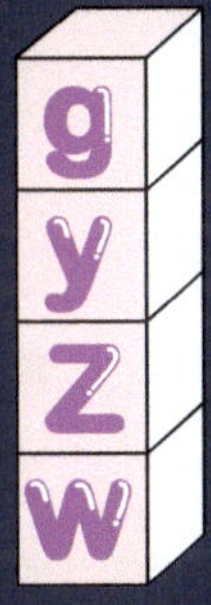

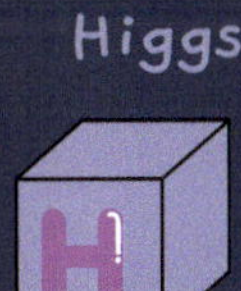

Elementary particles can mix and join.

New terms we must coin.

Baryons (protons and neutrons) get three quarks.

Mesons get a quark and an antiquark.

If you are confused.

What's what?

Who's who?

It's like a zoo.

All forces need a carrier.
Nature has selected that barrier.
Electromagnetic force (light)
needs photons.
Weak nuclear force
needs W and Z bosons.
Strong nuclear force
needs gluons.

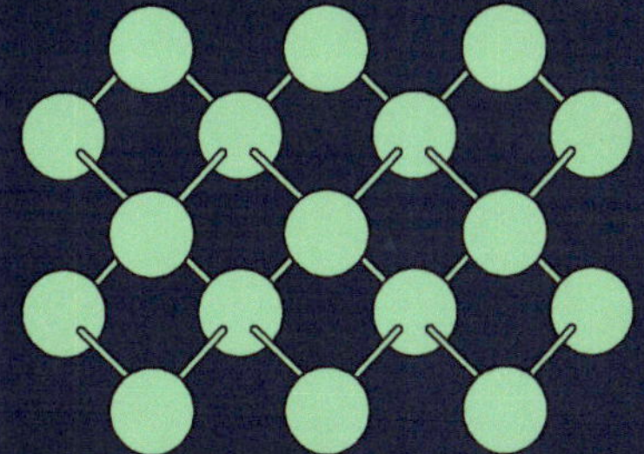 

Particles come in two forms,
fermions and bosons.
Fermions form matter
and bosons carry force.
Nature has chosen this course.

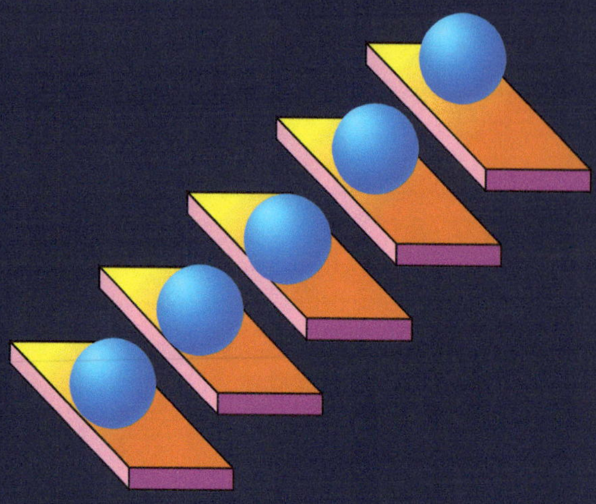

The difference is not only in their name.
Fermions and bosons don't behave the same.
Fermions(electrons) don't like each other one bit.
In the same quantum state, they can't sit.

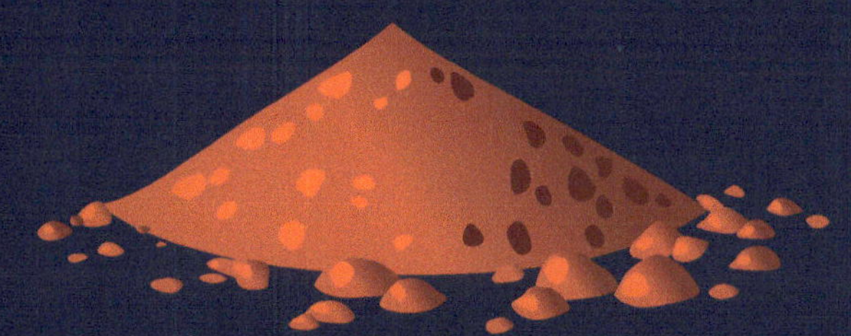

Bosons(photons) don't care where the others are at.
They can pile on each other just like that.

What happens in a nuclear test?
An atom has a lot of energy even at rest.

Mass gets converted into energy.
That's what Einstein told.
His equation is simple yet bold.

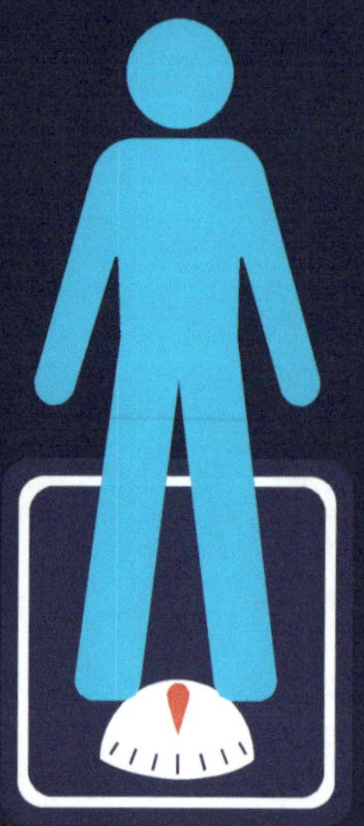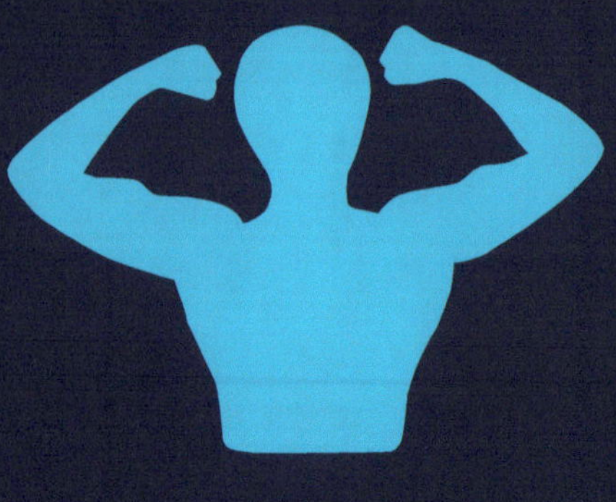

Energy and mass are the same.
The difference is only in their name.

Let's look at a matter
that looms large.
The antimatter is the same
as matter except for the opposite charge.

Electron's antimatter partner
is the positron.
In the early universe, antimatter disappeared
and the matter somehow won.

For experiments, particle accelerators
are the places to be.
Where particles collide
at great speed,
and there's much to see.

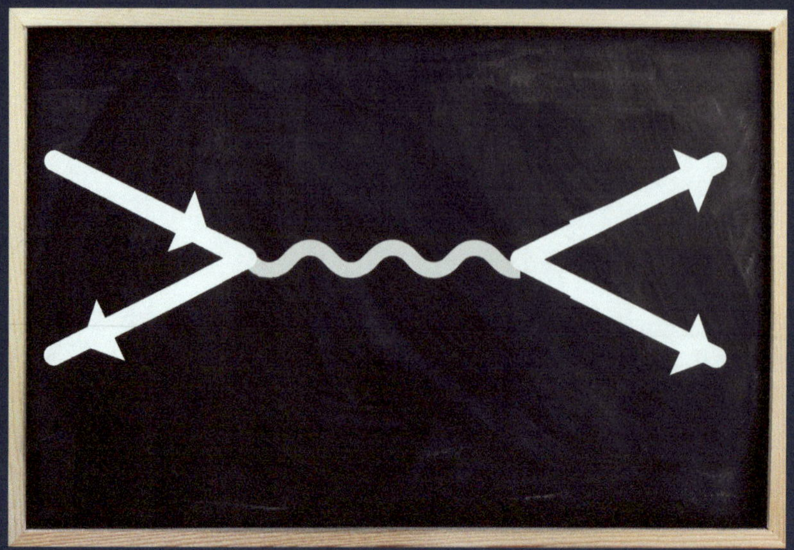

A picture is worth a thousand words
or maybe it's the other way around.

Feynman diagrams tell us how,
why, and where particles are found.

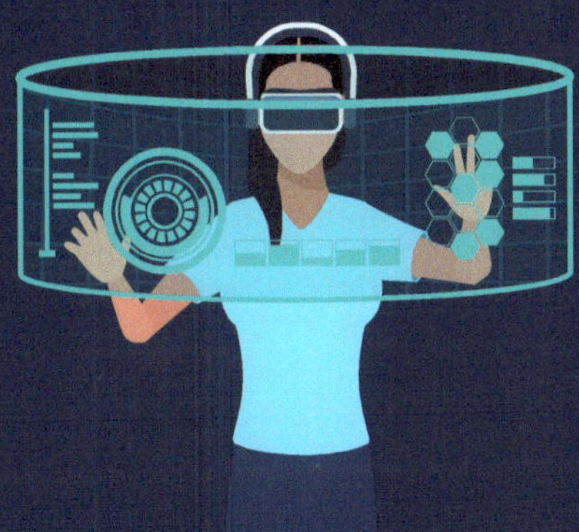

There's another duality.
There exists a virtual reality.

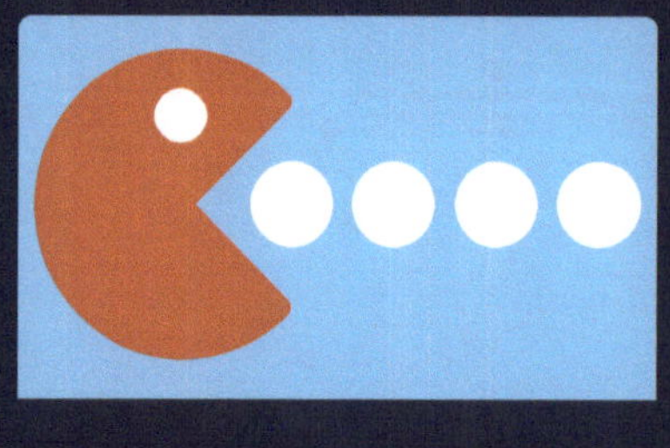

Virtual particles don't exist
for too long.
In the real world,
they don't belong.
Only in theory do they exist.
Perturbation theory
makes that list.

In the universe gravity reigns,
but inside the atom, its absence remains.

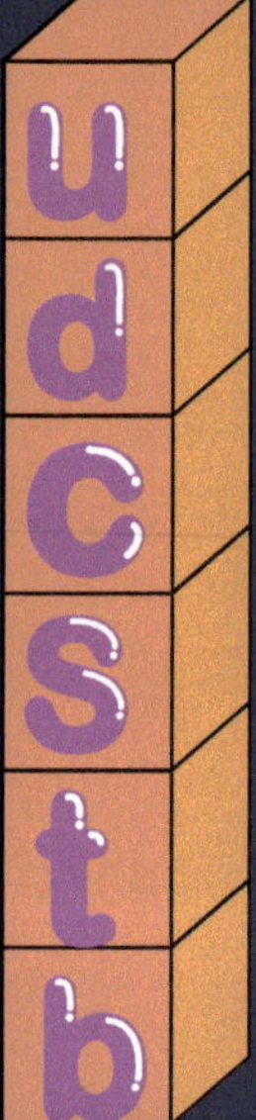

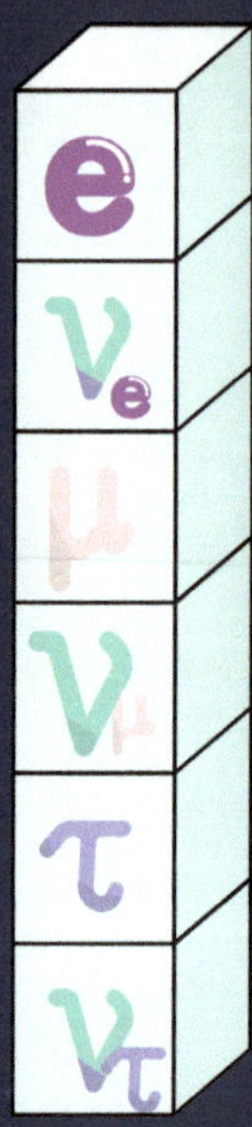

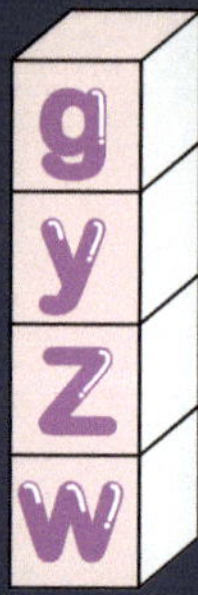

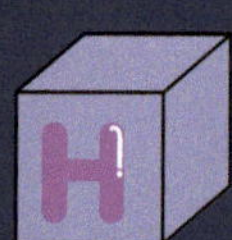

Many scientists already suspect
that without gravity,
the standard model cannot be correct.

If the standard model we lose,
what else is there to choose?

Scientists have tried all sorts of stuff.
But the evidence is not enough.
Loop and string theories
have lost their shine.
Maybe some exciting
theory is just down the line.

About the Author

Preetinder Rahil
writes fiction, non-
fiction, and poems
that rhyme. He tries
to keep things
simple, fun, and
worth your time.